OUR UNIVERSE

THE Milky Way

**BY MARION DANE BAUER
ILLUSTRATED BY JOHN WALLACE**

Ready-to-Read

Simon Spotlight
New York   Amsterdam/Antwerp   London
Toronto   Sydney/Melbourne   New Delhi

In memory of my brother, Willis Dane—M. D. B.

For Kelsey—J. W.

SIMON SPOTLIGHT
An imprint of Simon & Schuster Children's Publishing Division
1230 Avenue of the Americas, New York, New York 10020
For more than 100 years, Simon & Schuster has championed authors and the stories they create. By respecting the copyright of an author's intellectual property, you enable Simon & Schuster and the author to continue publishing exceptional books for years to come. We thank you for supporting the author's copyright by purchasing an authorized edition of this book.
No amount of this book may be reproduced or stored in any format, nor may it be uploaded to any website, database, language-learning model, or other repository, retrieval, or artificial intelligence system without express permission. All rights reserved. Inquiries may be directed to Simon & Schuster, 1230 Avenue of the Americas, New York, NY 10020 or permissions@simonandschuster.com.
This Simon Spotlight edition September 2025
Text © 2025 by Marion Dane Bauer
Illustrations © 2025 by John Wallace
All rights reserved, including the right of reproduction in whole or in part in any form. SIMON SPOTLIGHT, READY-TO-READ, and colophon are registered trademarks of Simon & Schuster, LLC. For information about special discounts for bulk purchases, please contact Simon & Schuster Special Sales at 1-866-506-1949 or business@simonandschuster.com.
Simon & Schuster strongly believes in freedom of expression and stands against censorship in all its forms. For more information, visit BooksBelong.com. The Simon & Schuster Speakers Bureau can bring authors to your live event. For more information or to book an event, contact the Simon & Schuster Speakers Bureau at 1-866-248-3049 or visit our website at www.simonspeakers.com.
Manufactured in the United States of America 0725 LAK
2 4 6 8 10 9 7 5 3 1
Library of Congress Cataloging-in-Publication Data
Names: Bauer, Marion Dane, author. | Wallace, John, 1966– illustrator. | Bauer, Marion Dane. Our universe.
Title: The Milky Way / by Marion Dane Bauer ; illustrated by John Wallace.
Description: Simon Spotlight paperback edition. | New York : Simon Spotlight, 2025. | Series: Our universe. Ready-to-read ; level 1 |
Summary: "Earth lies within the Solar System that is part of a spiral galaxy. Look up and see all that there is in the galaxy we live in, the Milky Way"—Provided by publisher.
Identifiers: LCCN 2024040204 (print) | LCCN 2024040205 (ebook) | ISBN 9781665959209 (hardcover) | ISBN 9781665959193 (paperback) | ISBN 9781665959216 (ebook)
Subjects: LCSH: Milky Way—Juvenile literature.
Classification: LCC QB857.7 .B38 2025 (print) | LCC QB857.7 (ebook) | DDC 523.1/13—dc23/eng/20250111
LC record available at https://lccn.loc.gov/2024040204
LC ebook record available at https://lccn.loc.gov/2024040205

Glossary

- **asteroids** (say: AH-stuh-royds): small, rocky bodies found in space that circle the Sun.
- **black hole**: a region in space with such a strong gravitational field that no light or matter can escape it.
- **comets** (say: KAH-mits): space objects made of ice that circle the Sun and often have long tails made of gas and dust.
- **disk**: a flat or thin object.
- **estimate** (say: ESS-tih-mate): to judge approximately.
- **galaxy** (say: GAH-luhk-see): a very large group of stars, planets, gas, dust, and other matter. Galaxies are found throughout the universe.
- **nebulae** (say: NEH-byuh-lee): clouds of gas and dust in outer space.

Note to readers: Some of these words may have more than one definition. The definitions above match how these words are used in this book.

Go far from city lights.
Look up at the night sky.
See that band of light?

That is the Milky Way.

The Milky Way is
a huge cluster of stars,
planets, gas, and dust
called a **galaxy**.

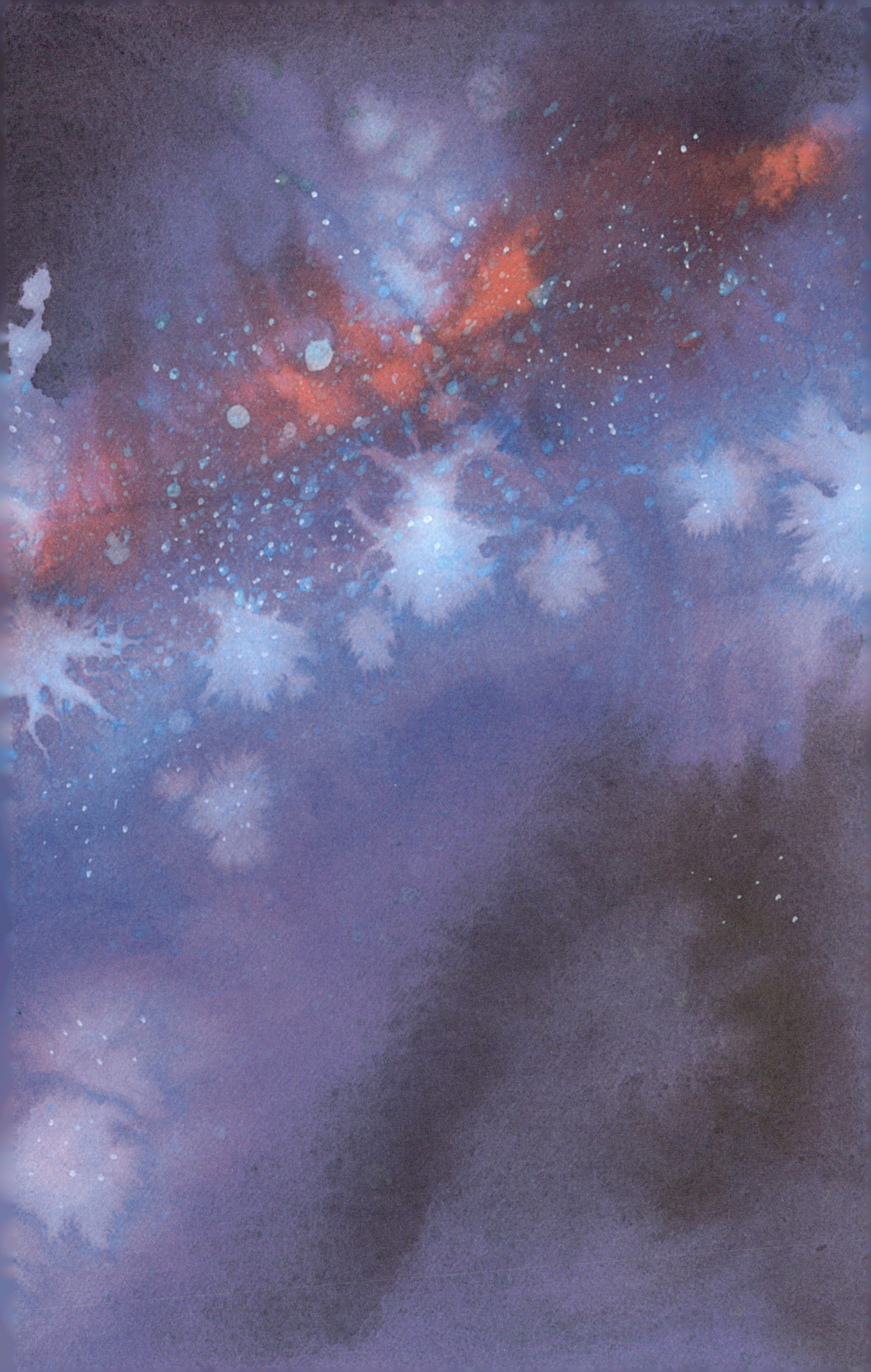

Every star you can see
with your bare eyes
is part of the Milky Way.

Our Sun is one
of hundreds of billions
of stars in our galaxy.

The Milky Way is a spiral galaxy.

From the side, it is shaped like a **disk**. From the top, it looks like a spinning pinwheel.

The Milky Way
has spiral arms.

Our solar system lies between two of those arms.

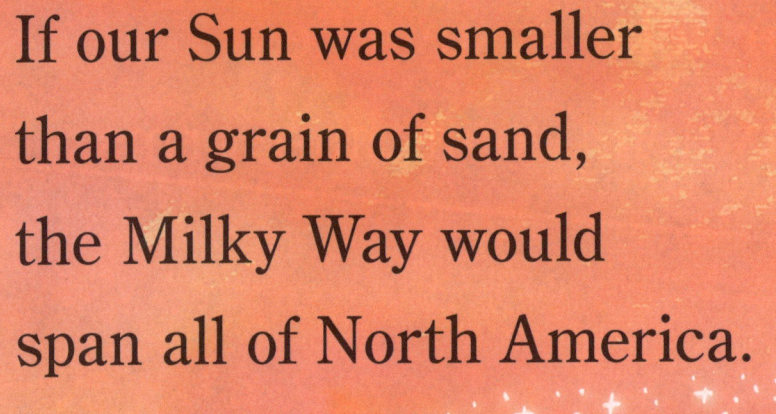

If our Sun was smaller than a grain of sand, the Milky Way would span all of North America.

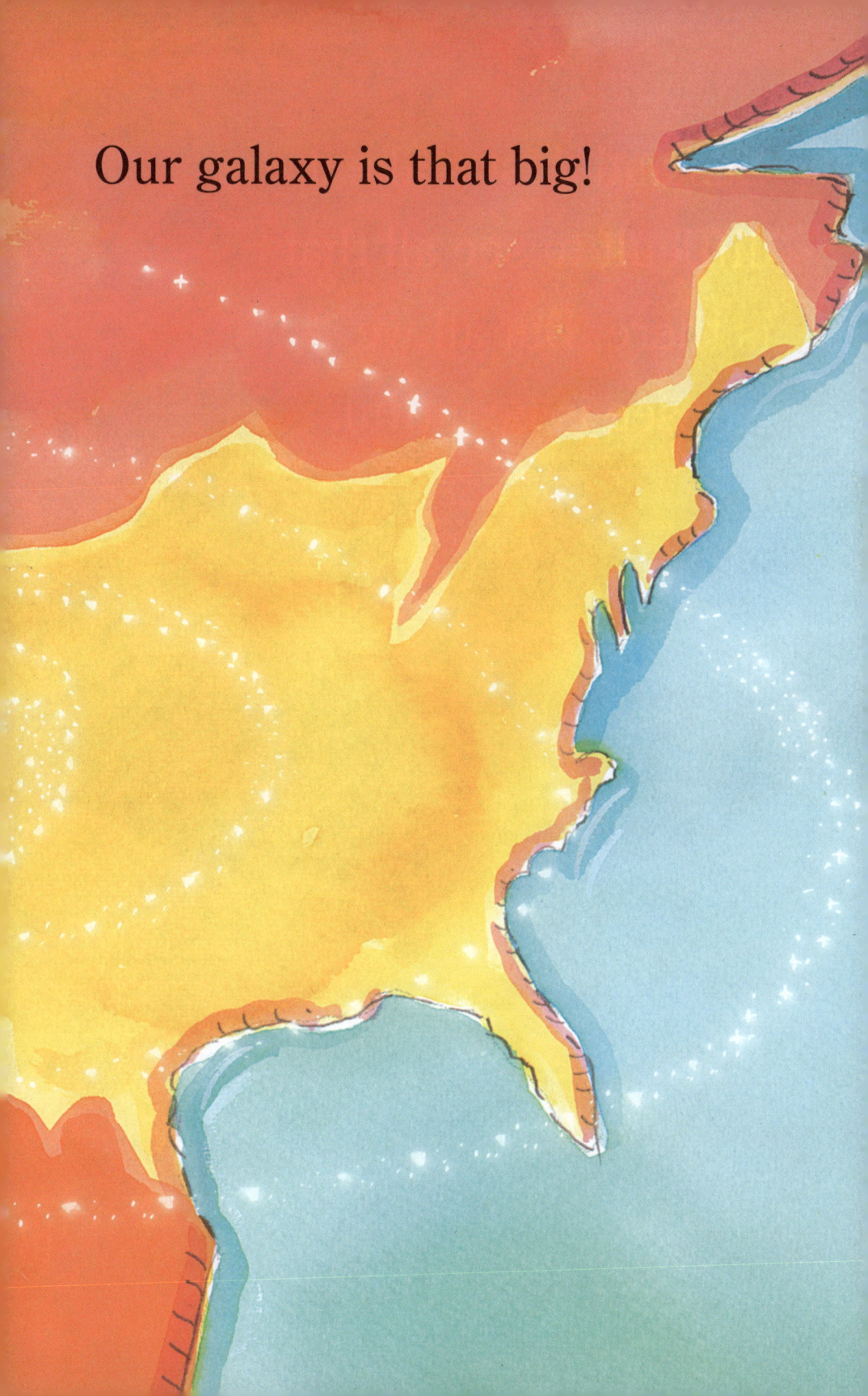

Our galaxy is that big!

Gas and dust make it difficult to see all that is there. But all we can see is beautiful.

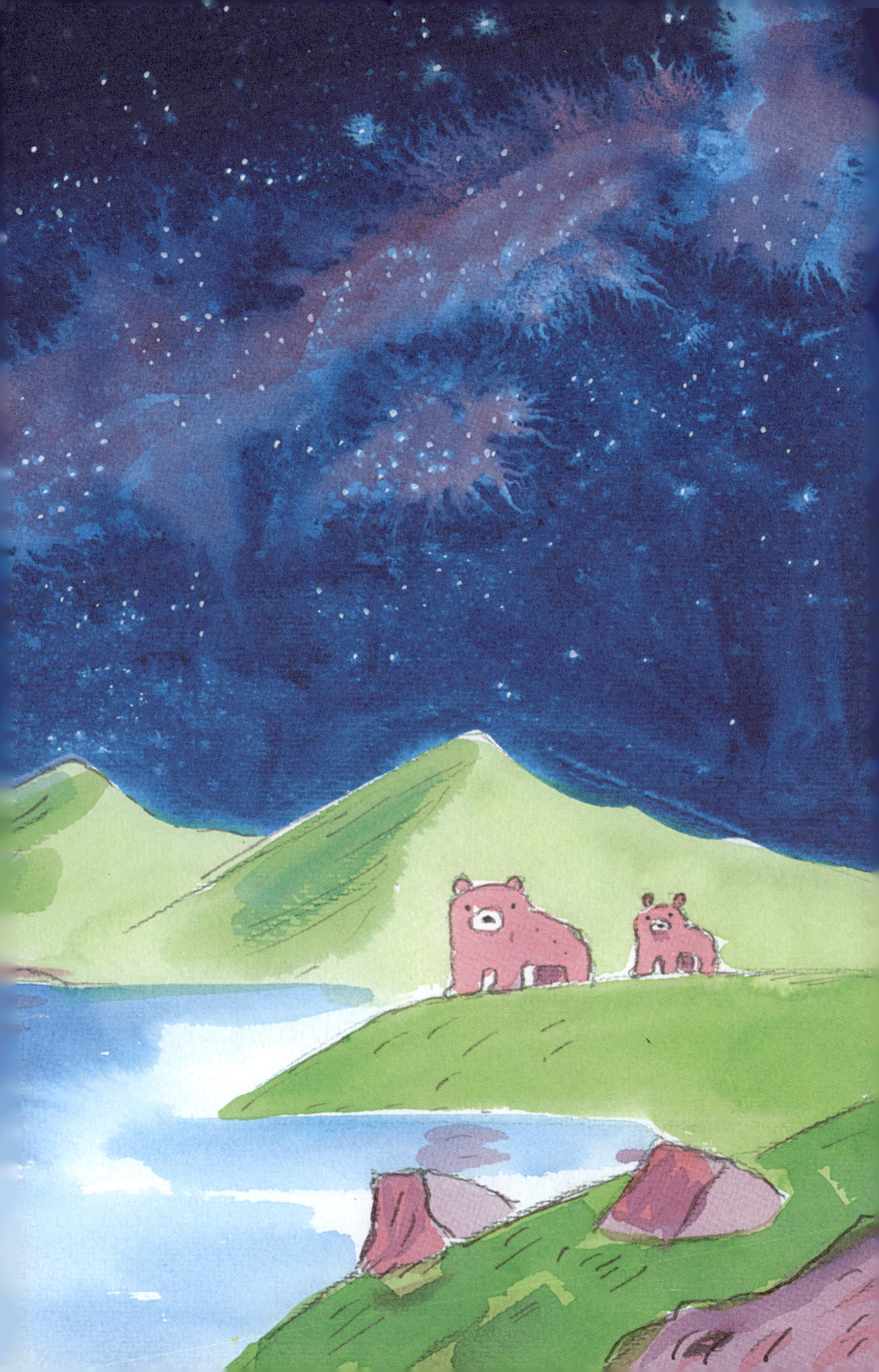

There are ancient stars
and colorful **nebulae**.
And a very quiet **black hole**
in the center too.

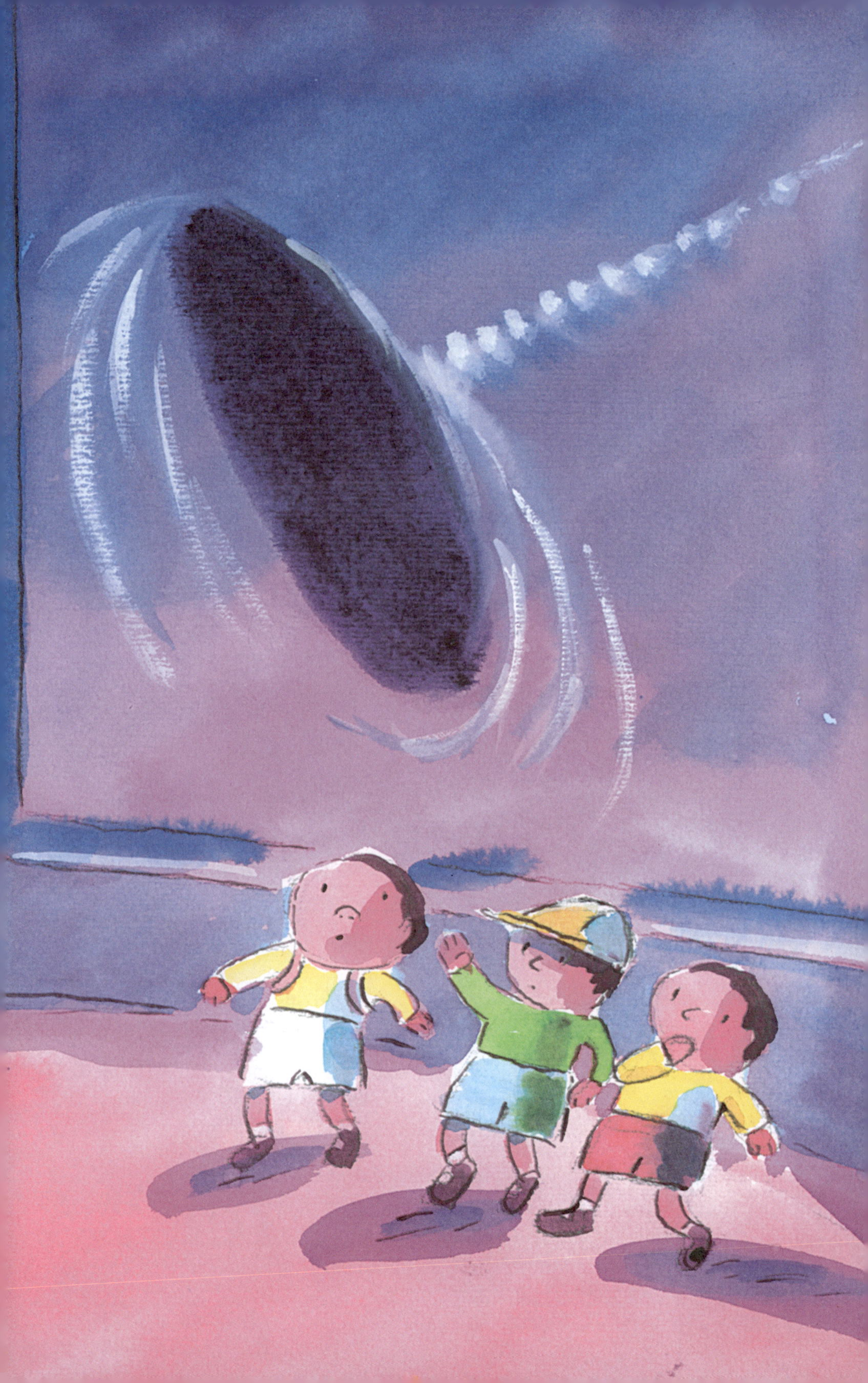

Scientists **estimate** that there are about two trillion more galaxies just in the part of the universe they can see.

And no one knows
how many more
we have not yet seen.

You live in
an amazing universe.

It is filled with stars,
and planets,
and moons.

And **asteroids**,
and **comets**,
and dust,
and gas.

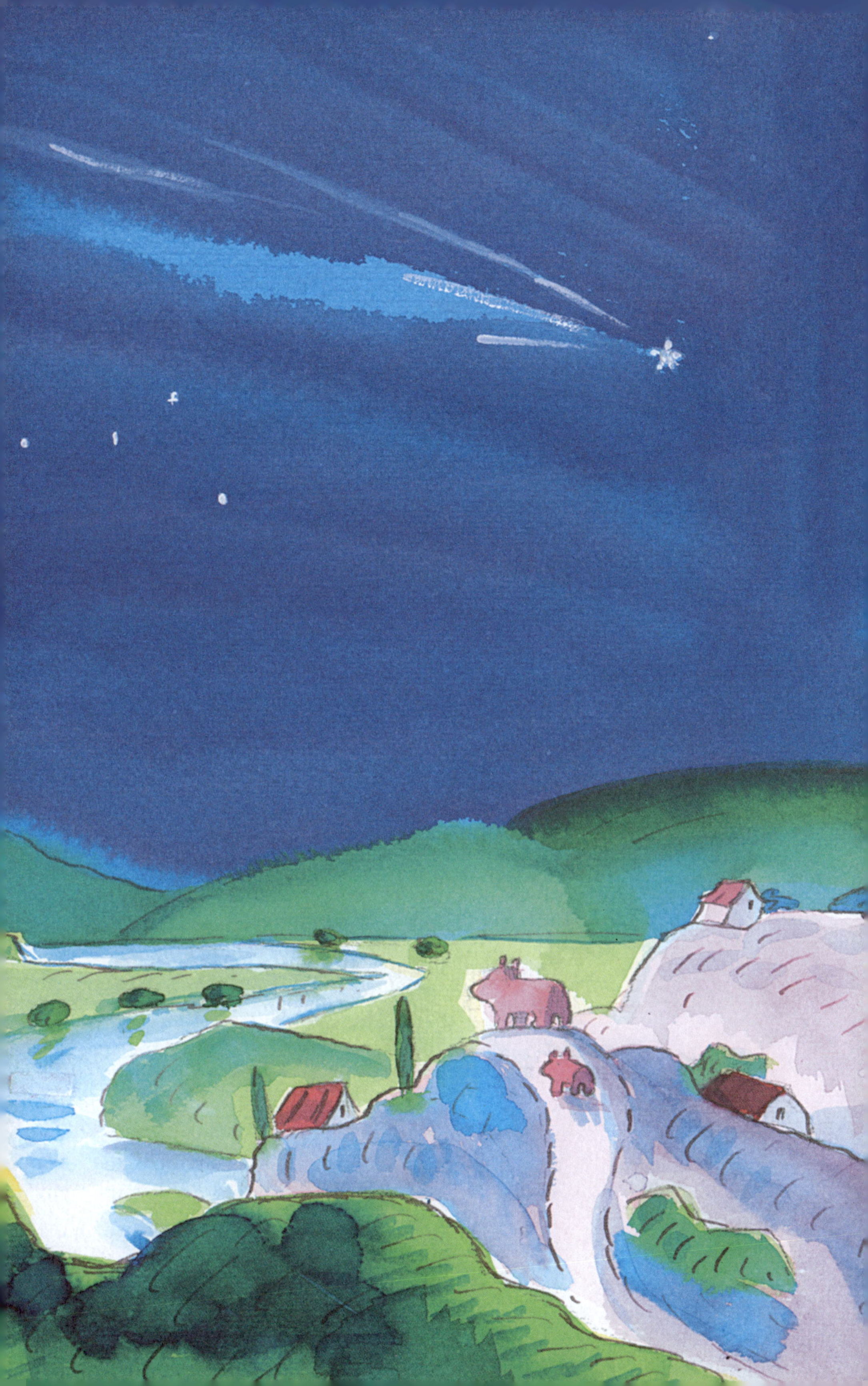

And lots, and lots,
and lots of space.

There is more empty space out there than anything else.

But down here on Earth there is you.

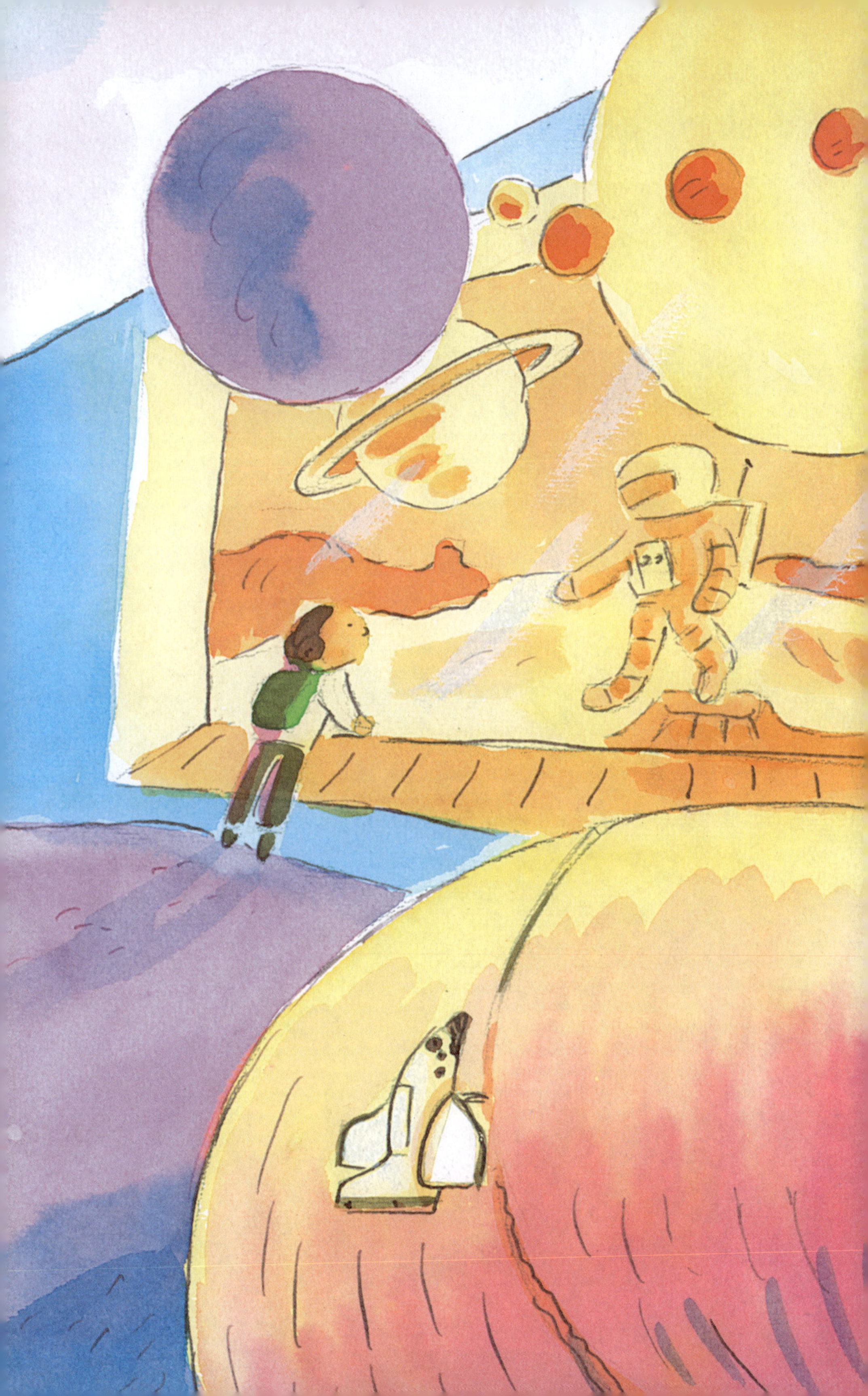

And all that sky!

Interesting Facts

+ Ancient Greeks told a story of the goddess Hera spraying milk across the sky. That is why we call our galaxy the Milky Way. In China that same band of stars is called the Silver River; in southern Africa it is called the Backbone of Night. It reminded hunter-gatherers in sub-Saharan Africa of campfire embers, while Polynesian sailors saw a cloud-eating shark.

+ Galaxies come in many sizes and shapes: spiral, elliptical (egg shaped), or irregular. Some have only a few million stars. Some have over a trillion.

+ Sometimes the Milky Way steals stars from other galaxies.

+ Traveling at the speed of light, 186,000 miles per second, it would take 25,000 years to reach the center of the Milky Way from Earth.

+ Our solar system orbits the center of the Milky Way once every 230 million years.

+ Without a telescope, you can see about 6,000 stars from Earth. For every star you can see, there are probably 20 million more you can't see because they are too far away or are blocked by clouds of cosmic dust.

+ Scientists study other galaxies from the outside to learn about ours. They study the Milky Way from the inside to learn about other galaxies.